THE CITY LIBRARY
SPRINGFIELD, (MA) CITY LIBRARY

S0-FBZ-136

DISCARDED BY
THE CITY LIBRARY

EMILY'S
PLACE FOR
CHILDREN

This Item Is A Gift
By Generous Donors to
The Springfield Library Foundation

NOV 0 3 2021

BIG JOBS, BIG TOOLS!

INCREDIBLE SUBMARINES

MARIE ROGERS

PowerKiDS press

New York

Published in 2022 by The Rosen Publishing Group, Inc.
29 East 21st Street, New York, NY 10010

Copyright © 2022 by The Rosen Publishing Group, Inc.

All rights reserved. No part of this book may be reproduced in any form without permission in writing from the publisher, except by a reviewer.

First Edition

Portions of this work were originally authored by Kenny Allen and published as *Submarines*. All new material in this edition authored by Marie Rogers.

Editor: Greg Roza
Cover Design: Michael Flynn
Interior Layout: Rachel Rising

Photo Credits: Cover, p.1 razihusin/iStock/Getty Images; pp. 4, 6, 8, 10, 12, 14,16, 18, 20, 21 13Imagery/Shutterstock.com; p. 5 S.Bachstroem/Shutterstock.com; p. 7 The Mariner 2392/Shutterstock.com; p. 9 noraismail/Shutterstock.com; p. 11 YASUYOSHI CHIBA/Staff/ADP/Getty Images; p. 13 Sergiy1975/Shutterstock.com; p. 15 Grenet Cedric/EyeEm/Getty Images; p. 17 U.S. Navy/Handout/Getty Images News/Getty Images; p. 19 Bettmann/Contributor/Getty Images; p. 21 Kuleshov Oleg/Shutterstock.com.

Library of Congress Cataloging-in-Publication Data

Names: Rogers, Marie, author.
Title: Incredible submarines / Marie Rogers.
Description: New York : PowerKids Press, [2022] | Series: Big jobs, big
   tools! | Includes bibliographical references and index.
Identifiers: LCCN 2020021708 | ISBN 9781725326750 (library binding) | ISBN
   9781725326736 (paperback) | ISBN 9781725326743 (6 pack)
Subjects: LCSH: Submarines (Ships)–Pictorial works–Juvenile literature. |
   Submarines (Ships)–Design and construction–Pictorial works–Juvenile
   literature.
Classification: LCC VM365 .R555 2022 | DDC 623.825/7–dc23
LC record available at https://lccn.loc.gov/2020021708

Manufactured in the United States of America

Some of the images in this book illustrate individuals who are models. The depictions do not imply actual situations or events.

CPSIA Compliance Information: Batch #CSPK22. For Further Information contact Rosen Publishing, New York, New York at 1-800-237-9932.

Find us on

CONTENTS

Under the Water 4
Parts of the Sub. 6
Diving Sub. 8
Working Closely Together. 10
Start the Engine! 12
The Fastest Sub 14
What Subs Do 16
In the Ohio Class 18
The Russian Typhoon 20
Inside the Typhoon Class 21
Glossary 22
For More Information 23
Index 24

Under the Water

Submarines, or subs, are boats that can go under the water. Some submarines are very small. The smallest subs don't even hold people. They're robots! Other subs are so big they can hold more than 150 people.

5

Parts of the Sub

Submarines are narrow on each end. They are also long and smooth. This allows subs to move easily though water. Many subs have a tower and a periscope. People inside the sub can use the periscope to look around.

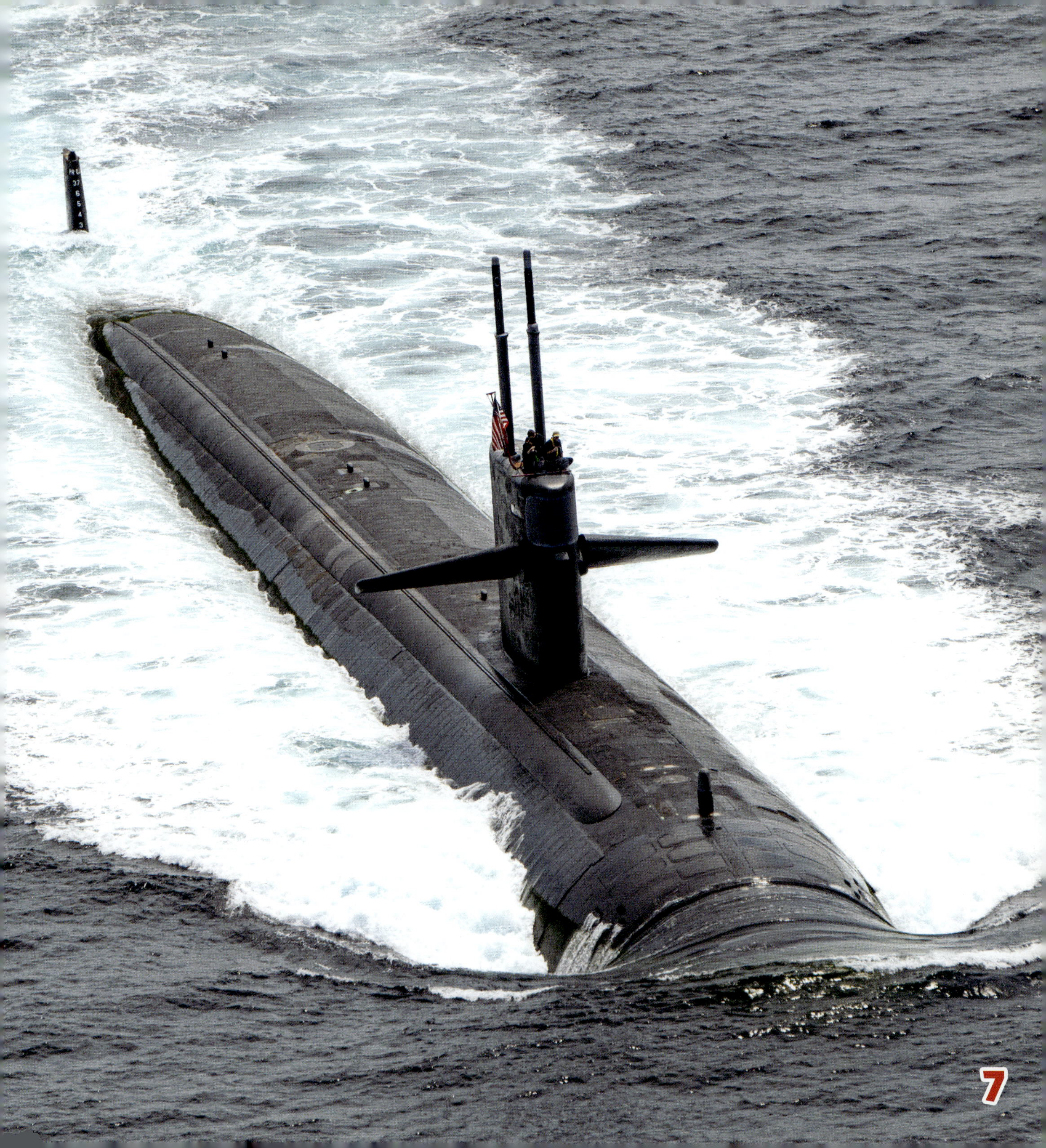

Diving Sub

Submarines need to dive under the water. They have tanks filled with air. When the sub needs to dive, the tanks are filled with water. This makes the sub heavier. Some subs can reach the ocean floor!

9

Working Closely Together

Workers on a submarine are called the crew. They don't have a lot of room to work in a sub. Crew members work for 6 hours and then have 12 hours of free time. They sleep in small beds called bunks.

periscope →

Start the Engine!

Submarines have **engines**. Engines create power and electricity. Some sub engines burn **fuel**, somewhat like the engine in a car. Some have steam engines. Many military submarines have engines called **nuclear reactors**.

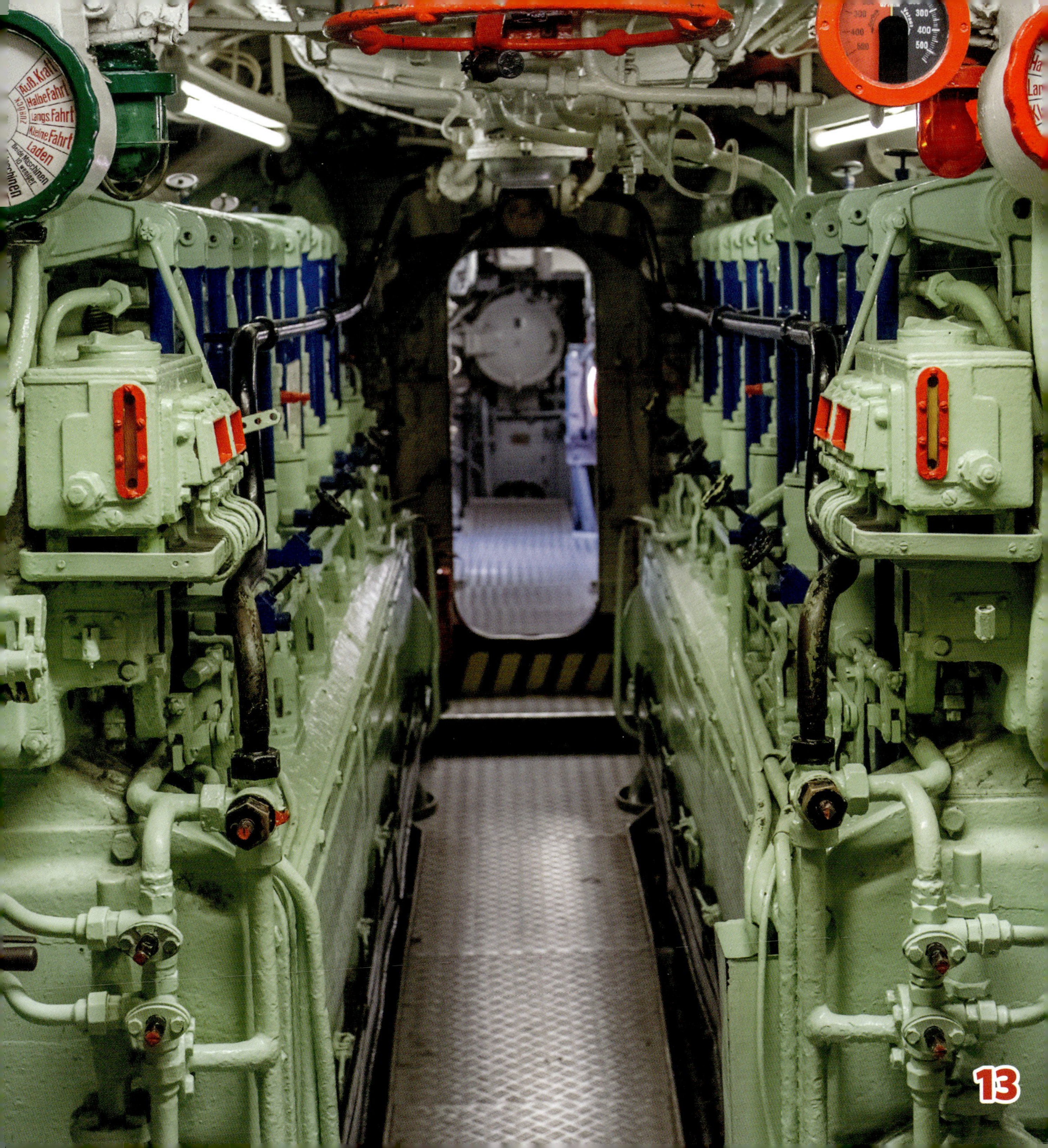
Halbe Fahrt
Langs. Fahrt
Kleine Fahrt
Laden

The Fastest Sub

A sub's engine turns giant propellers. Propellers have fins that spin in the water. This makes the sub move forward or backward. The Russian K-222 submarine once reached a speed of 51 miles (82 km) per hour!

15

What Subs Do

Most submarines are used by the military. They carry many weapons, such as **missiles**. Underwater missiles are called torpedoes. Other submarines are used to **explore** areas underwater. Some are used to fix **oil platforms**.

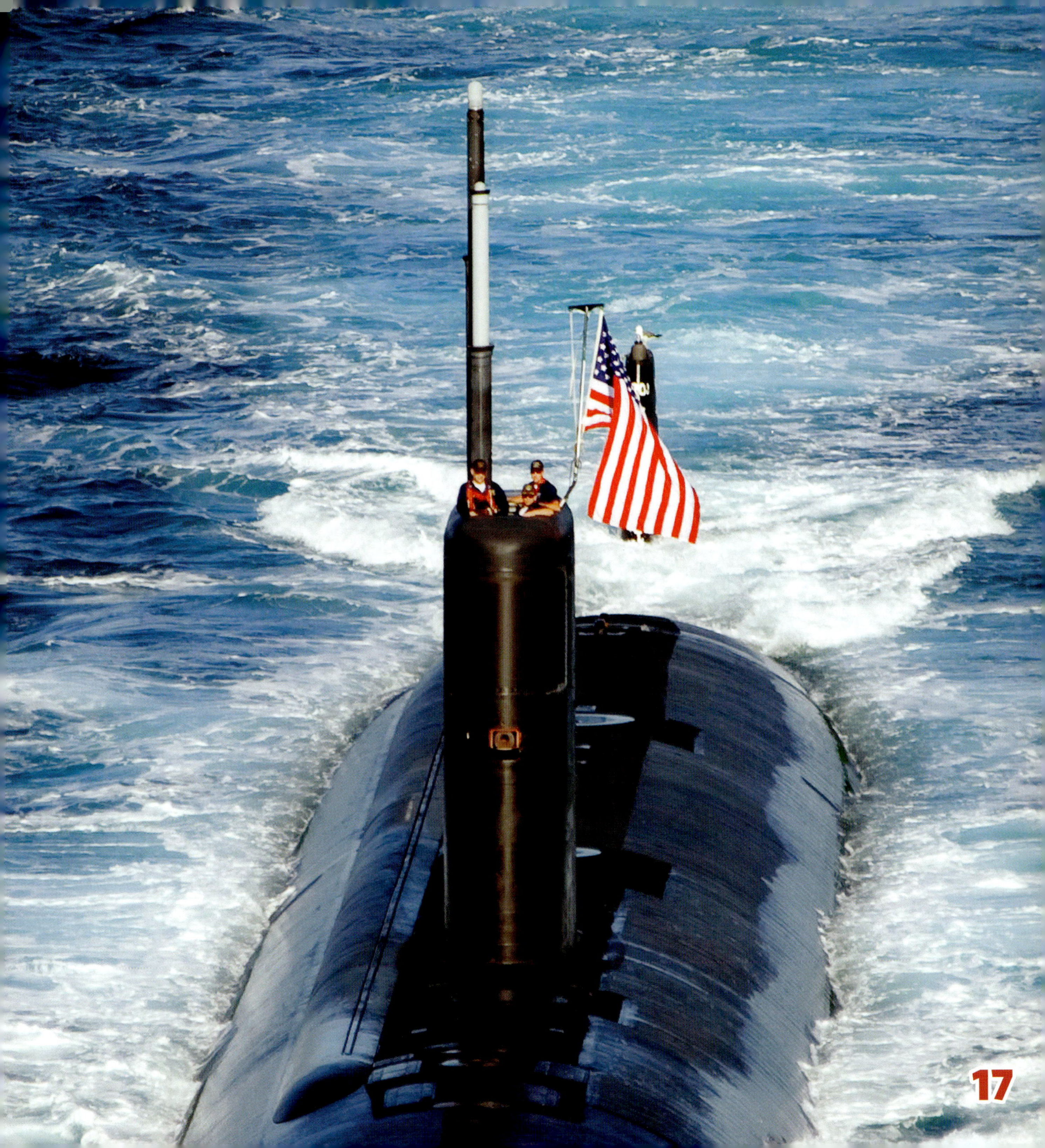

In the Ohio Class

The largest subs in the U.S. military are in the Ohio class, or kind. The first of these was the USS *Ohio*. It's been in use since 1981. There's a total of 18 Ohio class subs in the U.S. military.

19

The Russian Typhoon

The largest submarines ever built are the Russian Typhoon class subs. They're 564.3 feet (172 m) long! They can dive 1,312 feet (400 m) underwater and stay there for up to 180 days. Typhoon subs can carry 160 people.

Inside the Typhoon Class

- The first Typhoon class sub, the TK-208, was finished in 1981.

- Work on a seventh Typhoon class sub was started but never finished.

- Only two Typhoon class subs are in use today.

- Each Typhoon class sub has two nuclear reactors and two propellers.

- Typhoon class subs were made to break thick ice in the North Atlantic Ocean.

GLOSSARY

engine: A machine that produces motion or power for doing work.

explore: To search an area in a careful way to learn more about it.

fuel: A source of energy for a machine.

missile: An object that's shot to strike something from a distance.

nuclear reactor: A power plant that uses tiny pieces of matter called atoms to make energy.

oil platform: An oil drilling rig at sea.

FOR MORE INFORMATION

WEBSITES

Facts About Submarines
www.scienceforkidsclub.com/submarines.html
This website provides more information about submarines and their history.

Submarine
www.kids.britannica.com/kids/article/submarine/390261
Learn more about submarines at Britannica Kids.

BOOKS

Brody, Walt. *How Submarines Work*. Minneapolis, MN: Lerner Publications, 2019.

Gibbons, Gail. *Exploring the Deep, Dark Sea*. New York, NY: Holiday House, 2020.

Publisher's note to parents and teachers: Our editors have reviewed the websites listed here to make sure they're suitable for students. However, websites may change frequently. Please note that students should always be supervised when they access the internet.

INDEX

B
bunks, 10

C
crew, 10

E
electricity, 12
engines, 12, 14

F
fuel, 12

K
K-222 submarine, 14

M
missiles, 16

N
nuclear reactor, 12, 21

O
Ohio, USS, 18
Ohio class, 18
oil platforms, 16

P
periscope, 6, 11
propeller, 14, 21

R
robots, 4

T
tanks, 8
TK-208, 21
torpedoes, 16
tower, 6
Typhoon class, 20, 21